Rare 'Ring of Fire' Eclipse Stuns Observers in the Americas"

Roland Peterson

Introduction

The title, "Rare 'Ring of Fire' Eclipse Stuns Observers in the Americas", encapsulates the awe and excitement surrounding the celestial event. Eclipses have captivated humanity for centuries, and the prospect of a 'Ring of Fire' eclipse appearing over the Americas is a momentous occasion. The term "Ring of Fire" conjures a vivid image of a luminous halo encircling the darkened moon, setting the stage for a captivating visual spectacle.

Eclipses, both scientifically and culturally, have held profound significance. They inspire wonder, stimulate scientific inquiry, and carry rich mythological and folklore associations. This eclipse promises to be a rare treat, further underscoring its allure and the anticipation it generates among observers in the Americas.

The title hints at the collective astonishment and amazement expected as communities, astronomers, and skywatchers gather to witness this extraordinary event, reaffirming the enduring connection between humanity and the wonders of the cosmos.

Table of Content

Chapter 1

1.0 Introduction

1.1 Defining the 'Ring of Fire' Eclipse

The 'Ring of Fire' eclipse, scientifically known as an annular solar eclipse, is a breathtaking celestial event that captivates both astronomers and casual skywatchers. It derives its popular name from the striking visual effect created during this type of eclipse. Unlike a total solar eclipse where the moon completely covers the sun, in an annular eclipse, the moon appears slightly smaller in the sky. As a result, it cannot entirely obscure the sun, leaving a luminous ring or "ring of fire" around the edges of the moon.

This phenomenon occurs when the moon is near its apogee, the farthest point from Earth in its elliptical orbit, making it appear smaller in the sky. The sun and the moon align almost perfectly, with the moon positioned in front of the sun but not covering it entirely. This alignment creates a dazzling halo of solar brilliance encircling the moon, producing a surreal and awe-inspiring spectacle.

While not as dramatic as a total eclipse in terms of darkness and the dramatic change in the environment, the 'Ring of Fire' eclipse has its own unique charm. It showcases the delicate balance of celestial mechanics and provides observers with an opportunity to witness the sun's corona, the outer atmosphere of the sun, as it peeks out from behind the moon. For astronomers and sky enthusiasts, a 'Ring of Fire' eclipse is a reminder of the wonders of our solar system and the beauty of cosmic alignments.

1.2 Significance of the Event

The 'Ring of Fire' eclipse holds profound significance in the realm of astronomy and for anyone fortunate enough to witness this celestial marvel. Its significance can be understood on several levels.

First and foremost, these eclipses are vital for scientific research. During an annular eclipse, the sun's outer atmosphere, known as the corona, becomes visible as a radiant ring surrounding the moon. This offers astronomers a unique opportunity to study the corona, which is typically obscured by the sun's intense brightness. These observations contribute to our understanding of solar activity and aid in the prediction of space weather events, which can impact communication systems and even power grids on Earth.

Furthermore, the 'Ring of Fire' eclipse is a powerful reminder of the precision and predictability of celestial mechanics. It showcases the harmonious dance of the Earth, moon, and sun, emphasizing the intricate nature of our solar system. This event offers a chance for both amateur and professional astronomers to engage with the wonders of the cosmos and share their passion for the universe with the public.

On a more cultural note, the 'Ring of Fire' eclipse has historical and mythological significance in various societies. Many cultures throughout history have attached meaning and stories to these celestial events, making them a source of cultural heritage and inspiration.

In essence, the 'Ring of Fire' eclipse is not only a visual spectacle but a window to scientific exploration, a testament to the cosmos' precision, and a cultural touchstone that connects humanity to the marvels of the universe.

1.3 The Americas as the Viewing Stage

The choice of the Americas as the viewing stage for the 'Ring of Fire' eclipse is a celestial gift to astronomers and skywatchers across the continents. This unique astronomical event graces North and South America, providing a remarkable opportunity for a diverse range of people to experience the wonder of the cosmos.

The vast expanse of the Americas offers a canvas for celestial spectacle. Stretching from the Arctic tundra in the north to the temperate and tropical regions in the south, this vast geographical diversity ensures that a wide array of observers can partake in the eclipse. From the urban skyscrapers of New York City to the remote deserts of the American Southwest or the lush rainforests of the Amazon, people in different landscapes can gather to witness this cosmic phenomenon.

Moreover, the accessibility and infrastructure across the Americas make it easier for astronomers and enthusiasts to plan their eclipse experience. Major cities, national parks, and open spaces provide ample viewing opportunities, while developed transportation networks facilitate travel to prime locations.

The Americas, with its rich cultural diversity, adds another layer of significance to the event. Communities across the continents come together to celebrate and share in the awe of a 'Ring of Fire' eclipse, reinforcing the sense of unity in the face of the grandeur of the universe.

Chapter 2

2.0 A Celestial Dance: Understanding Solar Eclipses

2.1 Solar Eclipse Basics

A solar eclipse is a remarkable celestial event that occurs when the moon passes between the Earth and the sun, temporarily blocking all or a portion of the sun's light. Understanding the fundamental principles of a solar eclipse is key to appreciating the nuances of an event like the 'Ring of Fire' eclipse.

There are three primary types of solar eclipses:

1. **Total Solar Eclipse**: In a total solar eclipse, the moon completely covers the sun, casting a shadow on the Earth and plunging the surrounding area

into darkness. This rare and awe-inspiring event reveals the sun's corona, a halo of hot, ionized gas, and allows for observations of the solar atmosphere.

2. Partial Solar Eclipse: During a partial solar eclipse, the moon partially covers the sun, leading to a crescent-shaped sun. This is a more common occurrence and can be observed from a broader geographic area.

3. Annular Solar Eclipse (Ring of Fire): In an annular eclipse, the moon covers the center of the sun, leaving a ring-like appearance, often referred to as the "ring of fire". This happens when the moon is near its apogee, appearing smaller and unable to fully obscure the sun.

The path of totality, where a total or annular eclipse is visible, is relatively narrow, while a partial eclipse can be observed over a broader region. Viewing a solar eclipse requires caution, as looking directly at

the sun can cause permanent eye damage. Special eclipse glasses or other safe viewing methods are essential for observing this phenomenon safely.

Solar eclipses, regardless of type, offer a chance to witness the extraordinary alignment of celestial bodies and serve as a reminder of the wonders of our solar system.

2.2   The   'Ring   of   Fire' Phenomenon

The 'Ring of Fire' phenomenon, also known as an annular solar eclipse, is a captivating celestial occurrence that never fails to leave a lasting impression on those fortunate enough to witness it. This phenomenon is rooted in the precise alignment of the Earth, moon, and sun, resulting in a dramatic and mesmerizing visual display.

At the heart of the 'Ring of Fire' eclipse is the moon's apparent size in the sky. When the moon is at or near its apogee, the farthest point from Earth in its elliptical orbit, it appears slightly smaller. In contrast, the sun remains nearly constant in size from our vantage point on Earth. When the moon passes directly between the Earth and the sun during this specific orbital configuration, it obscures the center of the sun, leaving the sun's outer edges exposed, creating a radiant ring or halo effect. This effect is the eponymous "ring of fire".

What makes the 'Ring of Fire' phenomenon so visually striking is this juxtaposition of the sun's blazing, fiery corona surrounding the dark disk of the moon. This eerie and beautiful sight draws not only astronomers but also sky enthusiasts and the curious alike. It stands as a testament to the precise choreography of celestial bodies, a celestial ballet that unfolds in the sky for a brief moment.

The 'Ring of Fire' phenomenon is a testament to the mathematical precision and awe-inspiring beauty of our solar system, serving as a reminder of the cosmic wonders that unfold above us and deepening our understanding of the universe.

2.3 How and Why Eclipses Occur

Eclipses, both solar and lunar, are some of the most fascinating and predictable celestial events. To comprehend how and why eclipses occur, one must delve into the mechanics of our solar system and the orbital dynamics of the Earth, moon, and sun.

Solar eclipses, which include the 'Ring of Fire' phenomenon, transpire when the moon passes between the Earth and the sun, casting a shadow on our planet. This alignment occurs during the new moon phase when the moon is positioned directly between the Earth and the sun. The sun's light is momentarily blocked, resulting in an eclipse.

To understand why solar eclipses don't happen every month, one must consider the moon's orbit. The moon's orbit around Earth is slightly inclined, which means it doesn't perfectly align with the Earth-sun plane. This inclination is roughly 5 degrees, and it's this tilt that prevents solar eclipses from occurring with every new moon. Solar eclipses only transpire when the new moon occurs near a lunar node, points in the moon's orbit where it intersects the Earth-sun plane. These nodal crossings create windows of opportunity for eclipses.

The 'Ring of Fire' phenomenon, or an annular eclipse, is a subset of solar eclipses that happens when the moon is near its apogee, the farthest point from Earth in its elliptical orbit. Because the moon is slightly smaller in the sky when it's at apogee, it doesn't completely cover the sun's disk during an eclipse. Instead, a brilliant ring of sunlight, the "ring of fire", encircles the moon. This unique occurrence is a result of the moon's apparent size being insufficient to entirely block the sun.

Understanding lunar eclipses requires a different perspective. Lunar eclipses occur during a full moon when the Earth comes between the sun and the moon. The Earth's shadow has two distinct parts: the penumbral shadow and the umbral shadow. The penumbral shadow is the outer region of Earth's shadow, and it causes a subtle dimming of the moon's surface. In contrast, the umbral shadow is the darker, central part of the shadow. When the moon passes through the Earth's umbral shadow, a total lunar eclipse occurs. The reddish hue often seen during a total lunar eclipse is a result of Earth's atmosphere refracting and scattering sunlight, casting a warm, coppery glow onto the moon.

Chapter 3

3.0 A Rare Cosmic Alignment

3.1 What Makes This Eclipse Rare

The 'Ring of Fire' eclipse, also known as an annular solar eclipse, is a relatively rare celestial event due to a combination of specific conditions in the orbital dance of the Earth, moon, and sun.

One of the primary factors contributing to the rarity of this phenomenon is the moon's elliptical orbit around Earth. The moon's path is not a perfect circle but an ellipse, which means its distance from Earth varies. During a 'Ring of Fire' eclipse, the moon is positioned at or near its apogee, the farthest point from Earth. At this distance, the moon appears smaller in the sky compared to its size during its perigee, the closest point to Earth. As a result, when the moon passes in front of the sun, it cannot completely block the sun's disk, leaving a radiant ring, or "ring of fire" around the edges.

Additionally, the moon's orbital plane is inclined to the Earth's orbital plane by about 5 degrees. This inclination leads to the existence of two nodes where the moon's orbit intersects the plane of Earth's orbit around the sun. Solar eclipses, including 'Ring of Fire' eclipses, only occur when a new moon closely aligns with one of these nodes. This specific alignment is a significant contributing factor to the rarity of annular eclipses.

The intricate combination of the moon's elliptical orbit, its size in the sky, and the requirement for precise alignment at a lunar node makes annular eclipses, such as the 'Ring of Fire' eclipse, infrequent and highly sought-after events for astronomers and sky enthusiasts. When these celestial conditions do come together, they offer a profound opportunity to witness the breathtaking beauty and intricate mechanics of our solar system in action.

3.2 Historical Context of 'Ring of Fire' Eclipses

Eclipses, including the mesmerizing 'Ring of Fire' phenomenon, have played a significant role in human history, culture, and scientific understanding. Throughout the ages, these celestial events have left a profound impact on various societies and have contributed to the development of astronomical knowledge.

One of the earliest recorded observations of solar eclipses dates back to ancient China, where they were seen as harbingers of change and were often associated with celestial omens. Early cultures in Mesopotamia, such as the Babylonians, also documented eclipses in cuneiform texts. These historical records demonstrate that eclipses were considered awe-inspiring events, often perceived as messages from the gods.

In ancient Greece, philosophers and astronomers like Anaxagoras and Thales began to offer more rational explanations for eclipses. Thales, for instance, is said to have predicted a solar eclipse in the 6th century BCE. Such advances laid the foundation for a more scientific understanding of these phenomena.

Eclipses also played a pivotal role in the development of celestial mechanics. Johannes Kepler's work on the laws of planetary motion in the 17th century, particularly his explanation of the elliptical orbits of celestial bodies, shed light on the underlying principles governing eclipses. Kepler's laws allowed astronomers to predict eclipses more accurately, enhancing our grasp of these events.

Moreover, the occurrence of 'Ring of Fire' eclipses has often ignited curiosity and fascination. These events, characterized by the moon not completely covering the sun, have been the subject of art, literature, and mythology in different cultures. They symbolize a unique cosmic alignment, where the fiery corona surrounding the darkened moon is a source of mystery and inspiration.

In contemporary times, 'Ring of Fire' eclipses continue to captivate the world. They offer not only scientific research opportunities but also opportunities for communities to come together to witness these celestial spectacles. Public interest in eclipses has led to educational initiatives, events, and the sharing of knowledge about these events.

In summary, the historical context of 'Ring of Fire' eclipses reflects humanity's evolving understanding of the cosmos, from early superstitions and myths to the development of scientific principles and a deep appreciation for the beauty of the universe.

These events continue to connect us with the celestial wonders that have fascinated people for millennia.

3.3 Predicting and Tracking the Event

The precision of predicting and tracking a 'Ring of Fire' eclipse exemplifies the remarkable advancements in modern astronomy. It's a testament to humanity's mastery of celestial mechanics and computational power.

Predicting when an eclipse will occur involves intricate calculations based on well-established laws of motion and gravitation. Astronomers use detailed models of the moon's orbit, Earth's orbit, and the positions of the sun and moon relative to Earth. With these data, they can forecast when the moon will pass in front of the sun, leading to a solar eclipse. For 'Ring of Fire' eclipses specifically, the moon's position in its elliptical orbit plays a crucial role in determining the occurrence of this phenomenon.

Highly accurate astronomical software, combined with centuries of observational data, helps astronomers make precise predictions. These predictions not only include the date and time of the eclipse but also the path of totality and regions where the 'Ring of Fire' effect will be visible.

To track the eclipse as it unfolds, astronomers and space agencies around the world use advanced equipment and techniques. Telescopes equipped with solar filters, specialized cameras, and even satellite-based observations are used to monitor the eclipse. These observations not only serve to document the event for scientific study but also provide valuable data for refining eclipse prediction models.

In recent years, technological advancements, such as high-resolution satellite imagery and the internet, have allowed people worldwide to track eclipses in real-time. Live webcasts and apps provide the public with the opportunity to witness the event from anywhere with an internet connection, further democratizing the experience.

The ability to predict and track 'Ring of Fire' eclipses showcases the immense progress in our understanding of the cosmos. This knowledge not only fosters scientific research but also allows people of all backgrounds to marvel at the celestial wonders above, strengthening the connection between humanity and the universe.

Chapter 4

4.0 Preparing for the Spectacle

4.1 Safety Precautions for Eclipse Viewing

Viewing a solar eclipse, including a 'Ring of Fire' eclipse, is an awe-inspiring experience, but it must be done with extreme caution to protect your eyes and vision. Here are essential safety precautions:

1. Use Certified Eclipse Glasses: Never look at the sun, even during an eclipse, without proper eye protection. Certified eclipse glasses with special solar filters provide safe viewing. Ensure they meet international safety standards to block harmful radiation.

2. Pinhole Projector: A safe way to indirectly view an eclipse is through a pinhole projector. This simple device projects the eclipse's image onto a surface, such as a piece of paper, allowing you to watch it without looking at the sun directly.

3. Solar Telescopes and Filters: If you plan to use telescopes or binoculars, make sure they have approved solar filters to cover the front aperture. This prevents concentrated sunlight from damaging your eyes.

4. No Sunglasses or Unfiltered Cameras: Regular sunglasses, even multiple pairs, do not offer sufficient protection. Similarly, unfiltered cameras and telescopes can damage your eyes or equipment.

5. Seek Shade: During the eclipse, consider standing in the shadow of a building or using a tree's shade. This indirect light is much safer to observe.

6. Online Streaming: If you're unable to acquire proper safety equipment, consider watching the eclipse through reputable live streams online. Many space agencies and observatories provide live broadcasts of celestial events.

7. Educate Others: Share eclipse safety information with family, friends, and the public. Ensure that everyone around you understands the risks associated with viewing an eclipse without proper eye protection.

Remember, even a small portion of the sun's surface exposed during an eclipse can cause permanent eye damage. Safeguard your vision and enjoy the 'Ring of Fire' eclipse safely by adhering to these precautions. Eclipse viewing should be a memorable experience, not one that leads to eye injury.

4.2 Ideal Viewing Locations in the Americas

Selecting the ideal location to witness a 'Ring of Fire' eclipse in the Americas requires considering several factors, including geographical features, climate, and accessibility. Here are some of the top locations for eclipse viewing:

1. **The American Southwest:** States such as Arizona, New Mexico, and Utah offer vast open spaces with clear skies, making them prime locations for eclipse viewing. The southwestern deserts and national parks provide unobstructed horizons.

2. **Texas Hill Country**: This region in Texas is known for its charming landscapes and relatively clear skies. It's a great spot to experience the 'Ring of Fire' eclipse in a picturesque setting.

3. **The Ozarks**: Located in Arkansas and Missouri, the Ozarks feature rolling hills and forests, which can provide a stunning backdrop for eclipse viewing. The elevation can enhance visibility.

4. **Northern and Western Canada**: For those in North America, locations in northern and western Canada can offer excellent vantage points. Areas in Alberta, Saskatchewan, and the Northwest Territories may provide clear skies and minimal light pollution.

5. **The Amazon Rainforest**: In South America, parts of the Amazon rainforest in countries like Brazil offer a unique experience. The lush jungle surroundings create an enchanting atmosphere for eclipse observers.

6. **Chile and Argentina**: These countries in South America have regions with favorable conditions for eclipse viewing. Chile's Atacama Desert and

Argentina's Patagonia region are popular choices due to their clear skies.

7. **Remote Islands**: Islands in the Caribbean, such as the Cayman Islands and the Turks and Caicos, can also be suitable locations for eclipse viewing due to their relatively low light pollution levels and tropical settings.

4.3 Photography and Equipment Tips

Capturing the 'Ring of Fire' eclipse through photography can be a rewarding and memorable experience. To ensure you get the best shots, it's essential to plan and equip yourself appropriately. Here are some photography and equipment tips:

1. **Use the Right Gear**:

   - Camera: Choose a DSLR or mirrorless camera with manual settings. These offer more control over exposure and focus.

- Lenses: A telephoto lens with a focal length of at least 200mm is ideal for capturing details. A wider lens can capture the surrounding landscape during totality.

- Tripod: Use a sturdy tripod to keep your camera steady during long exposures.

2. **Solar Filters**:

- Never photograph the sun without proper solar filters. Use a solar filter designed for camera lenses to protect your camera's sensor and your eyes. It should be certified for safe solar viewing.

3. **Manual Settings:**

- Shoot in manual mode to control exposure settings. Start with ISO 100 or lower, a fast shutter speed, and a mid-range aperture (f/8 to f/11).

- Use spot metering to measure the sun's brightness and adjust exposure settings accordingly.

4. **Focusing**:

   - Focus your camera manually. Use live view to magnify the sun and achieve precise focus. Once focused, turn off auto-focus to avoid refocusing during shooting.

5. **Bracket Your Shots**:

   - Capture a series of bracketed shots with different exposures to ensure you get the best image. This allows you to merge photos in post-processing for optimal results.

6. **Remote Shutter Release**:

   - Use a remote shutter release or a timer to minimize camera shake when taking the shot.

7. **Location Scouting**:

   - Visit your chosen location in advance to assess angles, check for potential obstructions, and plan your composition. This can be crucial for capturing a dramatic landscape along with the eclipse.

8. **Practice Beforehand**:

   - Familiarize yourself with your camera settings and practice shooting the sun or moon before the eclipse day. This will help you troubleshoot any issues.

9. **Capture the Eclipse Phases**:

   - Take shots throughout the eclipse phases, from the initial contact to the maximum eclipse. This will provide a sequence of images to document the event.

10. **Stay Flexible**:

   - Weather conditions can change, so be prepared to adjust your plans. Have a backup location and a flexible shooting schedule.

Remember that the safety of your eyes and your equipment is paramount. Always use proper solar filters, avoid looking directly at the sun through your camera's viewfinder, and take care to protect your camera from the intense sunlight. With the right equipment and a well-thought-out plan, you can capture stunning images of the 'Ring of Fire' eclipse and preserve the memory of this spectacular event.

Chapter 5

5.0 The Big Day: Observing the 'Ring of Fire' Eclipse

5.1 Eclipse Timing and Phases

Understanding the timing and phases of a 'Ring of Fire' eclipse is crucial for those who wish to experience this celestial event to the fullest. These phases provide a roadmap for what to expect during the eclipse, making it an even more immersive experience.

1. **First Contact (Partial Eclipse Begins):** The eclipse begins with the first contact. At this point, the moon's silhouette first becomes visible on the sun's disk. You'll notice a small "bite" taken out of the sun. This is the onset of the partial eclipse.

2. **Partial Phase**: As the eclipse progresses, the moon continues to move across the sun's face. More of the sun's disk becomes obscured, gradually darkening the sky. Observers will see the characteristic crescent shape of the sun.

3. **Maximum Eclipse (Ring of Fire):** The central and most spectacular phase is the maximum eclipse. Here, the moon is positioned perfectly at the center of the sun, but it doesn't fully cover the sun's disk. This creates the iconic "ring of fire" effect, with a luminous halo around the darkened moon. This phase is often short-lived and offers a breathtaking view.

4. **Partial Phase (Again):** After the maximum eclipse, the moon starts to move away from the center of the sun. The crescent shape of the sun reappears as more of its disk becomes visible.

5. **Last Contact (Partial Eclipse Ends):** The eclipse concludes with the last contact. At this point, the moon's silhouette no longer covers the sun's disk, marking the end of the partial eclipse.

6. **Post-Eclipse**: After the eclipse, the sky gradually returns to its normal brightness, and daylight is restored. Depending on your location, the entire eclipse experience, from the first to the last contact, can last several hours.

5.2 Real-time Experiences from Observers

The beauty of a 'Ring of Fire' eclipse lies not only in its celestial mechanics but also in the real-time experiences and emotions it evokes among observers. People from all walks of life, amateur astronomers to casual skywatchers, share their awe-inspiring experiences during these events.

As the eclipse begins, there's an unmistakable sense of anticipation. Observers gather in prime viewing locations, often in groups or at organized eclipse-watching events. The first contact, when the moon takes its initial "bite" out of the sun, is met with excitement and wonder. It marks the beginning of the eclipse journey.

During the partial phases, as more of the sun's disk becomes obscured, the atmosphere takes on an eerie, twilight-like quality. Observers note the temperature drop and the changing colors of the landscape, creating a sense of enchantment.

The moment of maximum eclipse, when the 'Ring of Fire' appears, is met with gasps of amazement and appreciation. The ring-shaped corona and the darkened moon produce an almost surreal sight. The sun's fiery halo becomes an otherworldly presence in the sky, captivating those fortunate enough to witness it.

For some, it's a moment of spiritual reflection, a reminder of the universe's grandeur. For others, it's a time to share their joy and amazement with fellow eclipse-watchers. Cameras click, and telescopes capture the moment, preserving the event for later reflection and sharing.

As the eclipse transitions from maximum to the final stages, there's a sense of shared experience and connection with the cosmos. Observers often describe feeling humbled and privileged to witness such a rare and mesmerizing event.

The return of daylight and the end of the eclipse evoke a sense of fulfillment and accomplishment. Observers share their stories and photos, strengthening the bond between humanity and the mysteries of the universe.

Real-time experiences from observers capture the essence of a 'Ring of Fire' eclipse, where science, wonder, and human connection converge in a brief and breathtaking moment of cosmic alignment.

5.3 Weather and Its Impact on Viewing

Weather plays a pivotal role in the experience of viewing a 'Ring of Fire' eclipse. Clear skies are essential for observers to witness this celestial event, but weather conditions can be unpredictable and, at times, disappointing. Here's an exploration of how weather impacts eclipse viewing:

1. **Clear Skies Are Paramount:** A 'Ring of Fire' eclipse requires clear, unobstructed skies for optimal viewing. Cloud cover can obscure the sun and, in turn, the eclipse, making it impossible for observers to see the event. Clouds, particularly thick or widespread ones, are the bane of eclipse watchers.

2. **Weather Forecasts and Timing**: Observers planning to witness the eclipse often rely on weather forecasts to choose their viewing location. These forecasts provide information about cloud cover, precipitation, and atmospheric conditions. It's essential to check weather updates leading up to the eclipse to make informed decisions regarding where to go.

3. **Impact of Local Climate**: Local climate conditions have a significant influence on eclipse viewing. Some regions, like deserts or high-altitude locations, are more likely to offer clear skies, making them popular choices for eclipse enthusiasts. Conversely, areas with frequent cloud cover, like rainforests, may pose greater challenges for viewing.

4. **Preparation for the Unpredictable:** Even in areas known for good weather, unexpected conditions can arise. It's essential for eclipse watchers to prepare for the possibility of

last-minute cloud cover or adverse weather. This might involve having a backup location in mind or being ready to relocate if necessary.

5. **Flexibility in Planning:** To increase the chances of seeing the eclipse, some observers plan to travel to multiple potential viewing locations or organize eclipse-chasing trips to places with historically clear skies during the eclipse season.

6. **Virtual Viewing Options**: In the age of technology, virtual viewing has become an option. Many organizations live-stream eclipse events, allowing people to watch from anywhere with an internet connection, mitigating the impact of unfavorable weather conditions at their location.

Chapter 6

6.0 Scientific Significance and Research Opportunities

6.1 Studying the Sun's Corona

A 'Ring of Fire' eclipse presents a unique opportunity for astronomers and researchers to study the sun's corona, its outermost layer. The corona is typically difficult to observe due to the sun's blinding brightness, but during a solar eclipse, this ethereal halo becomes visible. Here's why studying the sun's corona is so significant:

1. **Understanding Solar Activity**: The sun's corona is not only stunning but also a crucial area for scientific investigation. It holds important clues about the sun's activity, such as solar flares and

coronal mass ejections. These phenomena can impact Earth's magnetic field and, in some cases, disrupt power grids and communication systems.

2. **Solar Physics**: The corona is hotter than the sun's surface, a paradox that scientists are still working to fully explain. During a 'Ring of Fire' eclipse, researchers can collect data and conduct experiments to better understand the corona's extreme temperatures, magnetic fields, and dynamic behavior.

3. **Improving Space Weather Predictions:** Gaining insights into the sun's corona can help improve space weather predictions. The corona is a source of solar wind, which, when it interacts with Earth's magnetosphere, can lead to geomagnetic storms. These storms can affect GPS systems, satellite communications, and even power distribution networks.

4. **Solar Eclipse Science:** Researchers use 'Ring of Fire' eclipses as a natural laboratory to study the corona. They deploy specialized equipment and instruments to capture images and data during these brief moments, which can yield valuable information for future research.

5. **Advancing Solar Science**: Observations of the sun's corona during solar eclipses have contributed to advancements in solar science. These findings influence our understanding of the sun's structure and activity, which is fundamental to understanding the universe.

The study of the sun's corona during eclipses is an interdisciplinary endeavor, bringing together astronomers, physicists, and space scientists. By harnessing the brief window of visibility provided by 'Ring of Fire' eclipses, researchers continue to peel back the layers of solar mysteries, deepening our understanding of our nearest star and its influence on our planet.

6.2   Scientific   Experiments During Eclipses

Eclipses, including the 'Ring of Fire' eclipse, provide a unique opportunity for conducting a wide range of scientific experiments and observations. These celestial events offer a controlled environment to study various phenomena, and researchers from various fields eagerly seize these opportunities. Here are some scientific experiments commonly conducted during eclipses:

1. **Corona Observations**: One of the primary objectives during solar eclipses is to study the sun's corona. Researchers employ specialized instruments, such as coronagraphs and spectrometers, to capture detailed images and data about the corona's temperature, composition, and magnetic fields. These observations contribute to our understanding of solar activity and space weather.

2. **Exoplanet Atmosphere Studies**: Eclipses are not limited to our solar system. Astronomers use lunar eclipses to study exoplanets, planets orbiting stars other than our sun. When an exoplanet passes behind its host star, its atmosphere can be analyzed by studying the changes in starlight, providing insights into the exoplanet's composition.

3. **Solar Chromosphere and Photosphere:** High-resolution imaging during eclipses allows scientists to study the sun's chromosphere and photosphere, the layers just below the corona. These observations help researchers better understand the sun's internal structure and magnetic activity.

4. **Gravity and General Relativity**: During a total solar eclipse, astronomers can conduct experiments related to general relativity. The sun's gravity can bend the path of light from distant stars, creating an apparent shift in their positions. These tests validate Einstein's theory of general relativity.

5. **Earth's Atmosphere**: Lunar eclipses provide a platform to study Earth's atmosphere. Scientists can analyze the Earth's atmosphere by observing how it scatters and filters sunlight as it passes through the planet's shadow.

6. **Eclipse-Caused Temperature Changes**: Researchers have studied temperature changes in the Earth's atmosphere and surface during eclipses. Understanding these changes contributes to climate and environmental research.

7. **Biological Observations**: Eclipses are not solely for astronomers. Biologists and ecologists have observed how animals and plants respond to sudden darkness and the abrupt temperature drop during eclipses, gaining insights into circadian rhythms and ecological behaviors.

6.3 Contributions to Astronomy and Space Science

Solar eclipses, including the 'Ring of Fire' eclipse, have made significant contributions to the field of astronomy and space science throughout history. These celestial events provide unique opportunities for research, observation, and experimentation, enriching our understanding of the universe. Here are some notable contributions:

1. **Confirmation of General Relativity**: One of the most famous contributions came in 1919 when a total solar eclipse provided the means to test Albert Einstein's theory of general relativity. Sir Arthur Eddington led an expedition to Principe and Sobral, where they photographed the stars near the sun during the eclipse. The apparent shift in star positions due to the sun's gravitational field confirmed Einstein's predictions, solidifying the theory of general relativity.

2. **Solar and Stellar Studies:** Solar eclipses have been instrumental in advancing our understanding of the sun's outer atmosphere, or corona. By blocking the sun's disk, researchers can observe the corona more clearly, providing insights into its temperature, magnetic fields, and other properties. These observations contribute to the field of solar physics.

3. **Exoplanet Discoveries**: Eclipses, including lunar eclipses, are used by astronomers to study exoplanets by analyzing the changes in starlight as exoplanets pass behind their host stars. This technique has enabled the discovery and characterization of many exoplanets, expanding our knowledge of planets beyond our solar system.

4. **Atmospheric and Environmental Research**: Lunar eclipses provide a unique opportunity to study Earth's atmosphere and its optical properties. Researchers have used these events to gather data on atmospheric composition,

pollution, and climate effects, contributing to environmental and atmospheric science.

5. **Stellar Photometry and Star Studies**: Eclipses have enabled precise photometry of stars, allowing astronomers to measure stellar parameters such as radii, temperatures, and luminosities. These measurements aid in the understanding of stellar evolution and the behavior of stars.

6. **Planetary Observations**: Eclipses can be used to study the planets in our solar system. When the moon passes in front of a planet during a lunar eclipse, astronomers can observe the planet's occultation, offering valuable data about its atmosphere and physical properties.

7. **Education and Public Engagement**: Eclipses capture the public's imagination and provide unique opportunities for science education and outreach. These events generate enthusiasm for

space science, prompting the public to engage with astronomy and related fields.

Chapter 7

7.0 Community and Cultural Impact

7.1 Eclipse Mythology and Folklore

Eclipses, including the 'Ring of Fire' eclipse, have been a source of fascination and wonder in cultures around the world for centuries. They often feature prominently in myths, folklore, and cultural traditions, reflecting both curiosity about the cosmos and a sense of the mysterious. Here are some examples of eclipse mythology and folklore from various cultures:

1. **Ancient China**: In ancient Chinese mythology, it was believed that solar eclipses occurred when a celestial dragon devoured the sun. To ward off the

dragon and ensure the sun's return, people would create loud noises and launch fireworks during an eclipse.

2. **Norse Mythology**: In Norse mythology, it was believed that solar eclipses were caused by the wolf Sköll chasing the sun across the sky. When Sköll temporarily caught the sun during an eclipse, it was seen as a sign of impending doom. To prevent this, Norse warriors would make noise and engage in various rituals.

3. **Hindu Beliefs**: In Hindu mythology, the demon Rahu is said to be responsible for eclipses. According to legend, Rahu tried to consume the elixir of immortality, but the sun and moon informed Lord Vishnu, who severed Rahu's head. Rahu's severed head became Ketu, and his vengeance results in eclipses.

4. **Native American Traditions**: Various Native American tribes have their own eclipse mythology.

For example, the Pomo people in California believed that during a solar eclipse, the sun and moon were temporarily dimmed as they engaged in a conjugal embrace.

5. **Aztec Beliefs**: The Aztecs believed that a solar eclipse was an ominous event, often associated with the god Huitzilopochtli. It was seen as a time of potential disaster and misfortune, and sacrifices were made to appease the gods.

6. **Korean Folklore**: In Korea, it was traditionally believed that solar eclipses occurred when mythical dogs attempted to steal the sun. People would engage in various rituals, such as making loud noises and banging on pots, to scare away the dogs.

Eclipse mythology and folklore reveal the deep connection between humans and the cosmos. These stories not only reflect ancient attempts to explain celestial phenomena but also demonstrate the cultural significance of eclipses as events that inspire awe, fear, and rituals. Today, while scientific understanding has largely dispelled these myths, the wonder and magic of eclipses persist, making them captivating events that transcend time and culture.

7.2 Events and Festivals Around the Americas

The 'Ring of Fire' eclipse and other celestial events often inspire communities to come together and celebrate the wonders of the cosmos. Across the Americas, various events and festivals are organized to mark these occasions. Here are some notable gatherings:

1. **Solar Eclipse Festivals**: Many regions across North and South America host solar eclipse festivals, especially during 'Ring of Fire' eclipses. These festivals offer a blend of science, culture, and art. They typically include solar viewing with

specialized telescopes and filters, educational programs, live music, and family-friendly activities.

2. **Astronomical Conventions**: Annual astronomical conventions and gatherings draw enthusiasts and professionals from the field. These events often coincide with notable eclipses, providing a platform for sharing research, engaging in discussions, and fostering a sense of community among astronomers and skywatchers.

3. **Observatory Open Houses**: Many observatories open their doors to the public during eclipses, allowing visitors to witness the event through powerful telescopes and participate in informative sessions led by astronomers and educators.

4. **Educational Workshops**: Educational institutions and science centers often host eclipse-themed workshops, lectures, and

demonstrations to inform the public about the science behind eclipses and their significance.

5. **Local Celebrations**: Communities in the path of an eclipse sometimes organize local celebrations, including parades, art exhibitions, and cultural events that highlight the eclipse's influence on their region's heritage.

6. **Online Virtual Events**: In the age of digital connectivity, online live streams and virtual events have become popular ways to engage a global audience during eclipses. Science organizations and space agencies broadcast live coverage, allowing people worldwide to participate.

These events and festivals not only enhance public awareness of eclipses but also create opportunities for people of all ages to explore the cosmos, deepen their knowledge of astronomy, and appreciate the beauty of the natural world. Whether one is a dedicated astronomer or simply curious about the universe, these gatherings offer a chance to share in the marvel of celestial phenomena like the 'Ring of Fire' eclipse.

7.3 Stories and Reactions from Local Communities

Solar eclipses, particularly a 'Ring of Fire' eclipse, have a profound impact on local communities that find themselves in the eclipse's path. These events elicit a wide range of reactions and stories that reflect the wonder and cultural significance of eclipses. Here are some stories and reactions from local communities:

1. **Community Gatherings**: Local communities often come together to witness eclipses as a collective experience. Parks, schools, and town squares become gathering places where people share stories, telescopes, and eclipse glasses. The

sense of unity and awe in these moments is palpable.

2. **Cultural Traditions**: In some regions, solar eclipses are steeped in cultural traditions and folklore. Communities with ancient eclipse-related myths and rituals may incorporate them into eclipse viewing events, adding a layer of cultural significance to the experience.

3. **Local Businesses and Tourism**: Eclipses provide a unique economic boost to communities in the eclipse path. Hotels, restaurants, and local businesses often see an influx of eclipse enthusiasts and tourists. In some cases, small towns become the center of attention as the world flocks to witness the celestial event.

4. **School Excitement**: Eclipse days are often marked with excitement in local schools. Students and teachers alike get the opportunity to learn about astronomy and witness a rare event. Schools

may organize special eclipse-related projects and activities, fostering a love for science.

5. **Safety Preparations:** Local authorities and organizations take safety seriously during eclipses. They often provide information about eclipse viewing safety and organize events to ensure the public watches safely. These efforts contribute to a smooth and secure eclipse experience.

6. **Historical Significance**: Communities with a history of hosting eclipses often share stories and anecdotes passed down through generations. These narratives highlight the eclipse's historical significance and its impact on the region.

7. **Environmental Observations**: Eclipses can lead to observations of changes in the environment. Local flora and fauna may respond to the sudden darkness, and scientists and community members may document these behaviors.

Eclipses, such as the 'Ring of Fire', evoke a sense of wonder and curiosity that transcends borders and brings diverse communities together. They serve as powerful reminders of the interconnectedness of humanity with the cosmos. The stories and reactions of local communities during these celestial events add depth to the experience, fostering a sense of unity and shared wonder under the shadow of the moon.

Chapter 8

8.0 Conclusion and Reflection

8.1 Recap of the 'Ring of Fire' Eclipse

The 'Ring of Fire' eclipse, a striking celestial event, captivated observers across the Americas, leaving an indelible mark on those fortunate enough to witness it. As the moon passed in front of the sun, this event created a dramatic spectacle, characterized by a luminous ring surrounding the darkened moon. Here's a recap of the significant aspects of the 'Ring of Fire' eclipse:

1. **Unique Visual Phenomenon**: The 'Ring of Fire' eclipse, scientifically known as an annular eclipse, differs from a total solar eclipse. Instead of

completely obscuring the sun, the moon leaves a brilliant ring, or annulus, around its edges. This creates a remarkable sight, often described as a "ring of fire".

2. **Path of Visibility**: The path of visibility for this particular eclipse spanned regions of North and South America. Observers within this path had the opportunity to witness the full 'Ring of Fire' effect, while those outside the path saw varying degrees of a partial eclipse.

3. **Cultural Significance**: Eclipses hold cultural significance in many societies. They have been a source of fascination, myth, and folklore for centuries, often associated with celestial deities, dragons, and wolves. These stories reflect the profound impact of eclipses on human imagination.

4. **Scientific Exploration**: For astronomers and researchers, the 'Ring of Fire' eclipse provided a unique opportunity to study the sun's corona, its

outermost layer. These observations contribute to our understanding of solar activity and space weather.

5. **Community Engagement**: Local communities, businesses, and educational institutions within the eclipse path actively engaged in eclipse-related activities. Public gatherings, scientific workshops, and cultural celebrations brought people together to share in the experience.

6. **Safety Precautions**: Safety was a paramount concern, with authorities and organizations emphasizing the importance of using certified eclipse glasses and safe viewing techniques to protect eyes while witnessing the event.

8.2 The Next Eclipse: What to Expect

Eclipses are cyclical, and after one has passed, anticipation often turns to the next celestial event on the horizon. For those who are enthusiastic about witnessing eclipses, knowing what to expect in the future is part of the ongoing fascination with these cosmic occurrences. Here's a glimpse of what to anticipate for the next eclipse:

1. **Location and Visibility:** The location and visibility of an eclipse can vary significantly. Depending on where you are, you might witness a total solar eclipse, partial eclipse, or another type, like an annular or hybrid eclipse. These characteristics will determine the level of darkness and the unique visual effects you can expect.

2. **Timing and Frequency**: Eclipses occur regularly but not at the same location each time. The timing and frequency of eclipses depend on the

Earth's orbital motions and lunar cycles. Different types of eclipses can occur at varying intervals, making each one a distinct event.

3. **Preparation and Planning**: Planning ahead is essential for eclipse enthusiasts. Securing the proper viewing equipment, understanding the eclipse's path, and choosing a suitable location are crucial. Communities and organizations often prepare events and gatherings to celebrate these celestial occasions.

4. **Scientific Opportunities**: Eclipses continue to offer invaluable scientific opportunities. Researchers and astronomers eagerly await the next eclipse to conduct experiments, gather data, and enhance our understanding of the cosmos.

5. **Public Engagement**: Eclipses also provide opportunities for public engagement and education. Science centers, observatories, and educational institutions often host eclipse-related programs and

activities to encourage people to learn and participate in these awe-inspiring events.

8.3 Inspiring a New Generation of Astronomers

Eclipses, such as the 'Ring of Fire' eclipse, play a pivotal role in inspiring the next generation of astronomers and space enthusiasts. These celestial events ignite a sense of wonder and curiosity about the cosmos, fostering a deep interest in astronomy and related fields. Here's how eclipses inspire and engage young minds:

1. **Accessible Wonder**: Eclipses are easily accessible astronomical phenomena that don't require high-powered telescopes or complex equipment to observe. They provide an entry point for young people to explore the universe firsthand.

2. **Hands-On Learning**: Eclipses offer an interactive learning experience. Students and educators can engage in hands-on activities, from creating pinhole projectors to understanding the mechanics of celestial alignments. This experiential learning sparks a passion for science.

3. **Classroom Engagement**: Schools often incorporate eclipse events into their curriculum. Teachers use eclipses to teach concepts in physics, astronomy, and space science, making these topics more engaging and relatable to students.

4. **Public Outreach Programs:** Astronomical organizations and science centers host public outreach programs and events during eclipses. These initiatives allow children and families to interact with professional astronomers, view the eclipse safely, and ask questions about the universe.

5. **Career Aspirations**: Witnessing a remarkable event like a solar eclipse can inspire young

individuals to consider careers in astronomy, astrophysics, or related scientific fields. The experience can plant the seeds for future astronomers and researchers.

6. **Community Engagement**: Eclipses often draw entire communities together for shared experiences. Parents, educators, and community leaders can collaborate to organize eclipse viewing events, creating a sense of wonder and unity.

Inspired by eclipses, young minds are motivated to explore the night sky, ask profound questions about the universe, and pursue careers in scientific disciplines. As these budding astronomers continue their journey, they will contribute to our collective understanding of the cosmos and shape the future of space exploration and discovery.

Appendices

Glossary of Eclipse Terminology

1. **Solar Eclipse**: This occurs when the moon passes between the Earth and the sun, partially or fully blocking the sun's light. It can result in partial, total, or annular eclipses.

2. **Lunar Eclipse**: A lunar eclipse transpires when the Earth moves between the sun and the moon, casting a shadow on the moon. This can lead to a partial or total eclipse.

3. **Total Solar Eclipse**: During a total solar eclipse, the moon completely obscures the sun, creating a brief period of darkness. The sun's corona is visible as a radiant halo around the moon's edges.

4. **Annular Eclipse**: An annular eclipse occurs when the moon is farther from the Earth, appearing

smaller in the sky. This results in the sun's disk not being fully covered, creating a ring-like appearance, often called the 'Ring of Fire'.

5. **Partial Eclipse**: In a partial eclipse, only a portion of the sun is blocked by the moon, leading to a crescent-shaped appearance. The sun remains partially visible.

6. **Totality**: Totality refers to the brief period during a total solar eclipse when the sun is completely hidden by the moon. This is the most captivating phase for observers.

7. **Eclipse Path**: The specific geographic route along which an eclipse is visible. It is determined by the alignment of the sun, moon, and Earth.

8. **Corona**: The sun's corona is the outermost layer of its atmosphere, visible as a radiant halo during a total solar eclipse. It is hotter and less dense than the sun's surface.

9. **Umbra and Penumbra**: The umbra is the central, darkest part of the moon's shadow, while the penumbra is the outer, lighter region. These terms are commonly used during lunar eclipses.

10. **Diamond Ring Effect**: This is a striking feature during the moments just before and after totality in a solar eclipse. It occurs when a single point of the sun's surface is still visible, resembling a dazzling diamond ring.

11. **Baily's Beads**: Baily's Beads are small points of sunlight that briefly appear along the moon's edge just before and after totality during a total solar eclipse. They are caused by sunlight passing through lunar valleys.

Additional Resources and References

For those interested in delving deeper into the fascinating world of eclipses and related astronomical topics, there are a wealth of resources and references available. These sources provide valuable information, educational materials, and opportunities for further exploration:

1. Astronomical Organizations: Organizations like NASA, the European Space Agency (ESA), and the International Astronomical Union (IAU) offer comprehensive online resources about eclipses, space missions, and the latest in astronomical research.

2. Educational Institutions: Many universities and colleges provide online courses, lectures, and research papers on astronomy and eclipses. These academic resources can be a valuable source of in-depth knowledge.

3. Astronomy Magazines: Popular science magazines such as "Sky & Telescope" and "Astronomy" cover a wide range of astronomical topics, including eclipse predictions, observing guides, and scientific discoveries.

4. Astronomy Apps: There are numerous astronomy apps available for mobile devices that offer real-time sky maps, eclipse predictions, and interactive tools for exploring the night sky.

5. Planetariums and Science Centers: Local planetariums and science centers often host educational programs, exhibitions, and telescope viewings related to eclipses and space science.

6. Online Forums and Communities: Joining online astronomy forums and social media groups can connect you with fellow enthusiasts, providing opportunities for discussion, knowledge sharing, and advice on eclipse observation.

7. Books and Publications: There are many excellent books on eclipses, written for both beginners and experts. These books cover eclipse history, mythology, science, and observational techniques.

8. Official Websites: National space agencies like NASA and ESA maintain official websites with detailed information on space missions, scientific research, and upcoming eclipse events.

9. Astronomical Observatories: Visiting an astronomical observatory or research center can provide opportunities for hands-on learning and interaction with astronomers and their equipment.

10. Historical References: Exploring historical texts and records can offer insights into how past civilizations interpreted eclipses and their significance in culture and science.

These resources and references are valuable tools for anyone seeking to expand their knowledge of eclipses, astronomy, and space science. They cater to all levels of interest and expertise, from novice skywatchers to seasoned astronomers and researchers. You're a seasoned astronomer or a first-time observer, this glossary helps demystify the language of eclipses.

Index for Quick Reference

For a quick and convenient reference, here's an index summarizing the key subtopics covered in this discussion about eclipses, including the 'Ring of Fire' eclipse:

1. Ring of Fire Eclipse: An introduction to the 'Ring of Fire' eclipse, explaining its unique characteristics and visual effects.

2. Table of Contents: A structured outline of the subtopics covered in this discussion for easy navigation.

3. Defining the 'Ring of Fire' Eclipse: An exploration of the characteristics and scientific definition of the 'Ring of Fire' eclipse.

4. Significance of the Event: A discussion of the cultural and scientific importance of eclipses and their role in inspiring wonder and curiosity.

5. The Americas as the Viewing Stage: Insights into why the Americas are often a prime location for eclipse viewing and how communities engage with these events.

6. Solar Eclipse Basics: Fundamental information about solar eclipses, their types, and how they occur.

7. The 'Ring of Fire' Phenomenon: A detailed explanation of the 'Ring of Fire' effect during an annular eclipse and how it differs from total solar eclipses.

8. How and Why Eclipses Occur: A comprehensive overview of the celestial mechanics and reasons behind solar and lunar eclipses.

9. What Makes This Eclipse Rare: An exploration of the factors that contribute to the rarity of 'Ring of Fire' eclipses.

10. Historical Context of 'Ring of Fire' Eclipses: An examination of the historical significance of eclipses and their influence on various cultures.

11. Predicting and Tracking the Event: Insights into the methods and technology used to predict and track eclipses.

12. Safety Precautions for Eclipse Viewing: Important safety guidelines for observing eclipses to protect one's eyes and health.

13. Ideal Viewing Locations in the Americas: Tips on selecting optimal viewing locations for eclipse enthusiasts in North and South America.

14. Photography and Equipment Tips: A guide for capturing stunning eclipse photographs and selecting the right equipment.

15. Eclipse Timing and Phases: Detailed information about the timing and phases of a solar eclipse, from first contact to totality or annularity.

16. Real-time Experiences from Observers: Personal accounts and experiences shared by observers during eclipse events.

17. Weather and Its Impact on Viewing: An exploration of how weather conditions can influence eclipse visibility.

18. Studying the Sun's Corona: The significance of studying the sun's corona during solar eclipses and the scientific insights gained.

19. Scientific Experiments During Eclipses: An overview of the various experiments and research conducted during eclipses.

20. Contributions to Astronomy and Space Science: How eclipses contribute to the advancement of scientific knowledge and understanding of the universe.

21. Eclipse Mythology and Folklore: Cultural and mythological stories and beliefs associated with eclipses from different societies.

22. Events and Festivals Around the Americas: Celebrations, gatherings, and festivals held to commemorate eclipses in the Americas.

23. Stories and Reactions from Local Communities: The impact of eclipses on local communities and their shared experiences.

24. Recap of the 'Ring of Fire' Eclipse: A summary of the key aspects of the 'Ring of Fire' eclipse discussed in this conversation.

25. The Next Eclipse: What to Expect: Anticipating and preparing for upcoming eclipse events, including what to expect.

26. Inspiring a New Generation of Astronomers: How eclipses inspire and engage young minds, fostering an interest in astronomy and space science.

27. Glossary of Eclipse Terminology: A comprehensive glossary of terms related to eclipses and astronomy.

28. Additional Resources and References: A guide to further sources of information and educational materials related to eclipses and astronomy.

www.ingramcontent.com/pod-product-compliance
Lightning Source LLC
Chambersburg PA
CBHW060958260726
48661CB00005B/1928